Great Scientific Questions and the Scientists Who Answered Them

HOW DO WE KNOW THE NATURE OF THE ATOM

NATALIE GOLDSTEIN

Great Scientific Questions and the Scientists Who Answered Them

HOW DO WE KNOW THE NATURE OF THE ATOM

THE ROSEN PUBLISHING GROUP, INC.
NEW YORK

Published in 2001 by The Rosen Publishing Group, Inc.
29 East 21st Street, New York, NY 10010

First Edition

Library of Congress Cataloging-in-Publication Data

Goldstein, Natalie.
The nature of the atom / Natalie Goldstein. — 1st ed.
p. cm. — (Great scientific questions and the scientists who answered them)
Includes bibliographical references and index.
ISBN 0-8239-3385-7
1. Atomic theory—History—Juvenile literature. [1. Atomic theory. 2. Atoms.] I. Title. II. Series.
QD461 .G625 2001
539.14—dc21

00-011541

Cover images: subatomic particles.

Manufactured in the United States of America

Contents

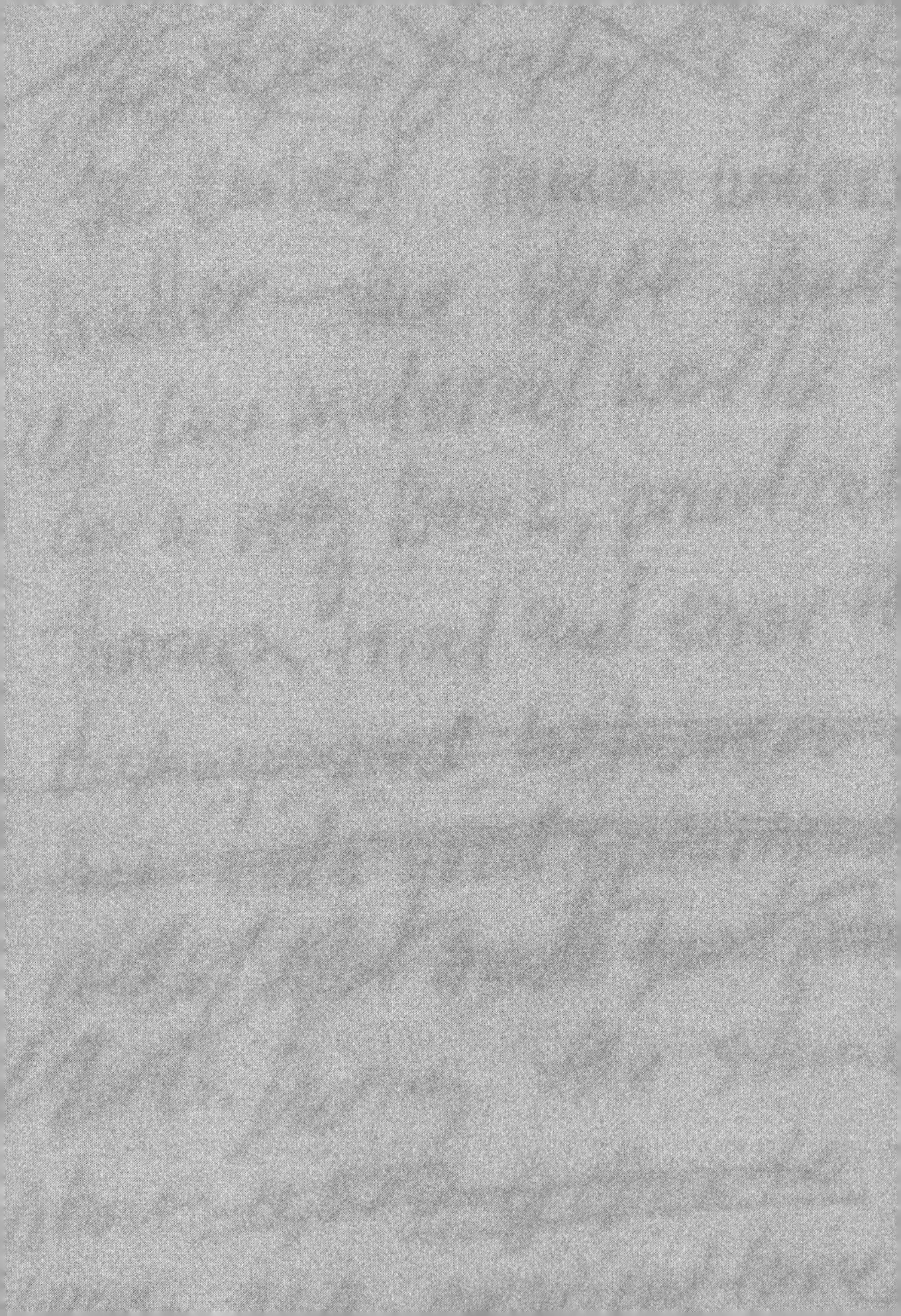

The Beginnings of Chemistry

The only things that exist are atoms and empty space. Everything else is mere opinion.

Democritus (ca. 400 BC)

The earliest humans understood matter—the stuff that makes up the material world—on a very basic, practical level. Through trial and

Humans learned to create and control fire 10,000 years ago.

error, they distinguished between rocks that made good spearheads and rocks that made good pounding stones. They were no doubt amazed by how matter was transformed into ashes when it burned.

During the Stone Age, about 10,000 years ago, humans learned to make and control fire. By the Neolithic period, about 6,000 years ago, agriculture and permanent human settlements were established. At this time, humans first discovered and began to use metals. No one is certain how the discovery was made. Experts suggest that small stones were placed around fires to contain them. When the fire burned itself out, people may have found brilliant nuggets of yellow gold or red copper—the first metals to be discovered—gleaming among the embers. People soon learned that heating certain rocks would yield these metals, and quite quickly they learned to shape and use them. By about 3,200 BC, food for the Egyptian pharaohs was cooked in copper pots. The Bronze Age began about 2,000 BC, when, no doubt by happy accident, rocks containing copper and rocks containing tin were heated together.

The combination of the two metals yielded the metal alloy bronze. The Hittites are credited with the discovery of iron around 1,500 BC. Metals played a significant role in the growth of chemical knowledge. It was necessary to learn how to separate valued metals from their ores, and as gold and silver coins began to be used as currency, it was important to learn how to determine if the metals were pure and of the proper weight.

Thales of Miletus suggested that all things were made of fundamental substances.

BASIC QUESTIONS

That pure metal could be separated from rock raised an interesting question. What were things really made of? What substances were "hidden" within other substances, and what were the most basic substances that constituted matter? Are gold and copper basic forms of matter? Or is there a more fundamental material that is common to both and to all other things? By 4,000 BC, Egyptians were heating combinations of sand, limestone, and soda (sodium carbonate) until the materials melted and turned into glass. The characteristics of glass are quite different from those of the materials used to make it. What was happening to material objects when they underwent such a dramatic transformation?

These questions intrigued the ancient Greeks, who were extremely curious people. Their greatest philosophers tackled the question of the nature of matter. Thales of Miletus, who lived around 600 BC, stated that all material things are aspects of one fundamental substance—water. Thales saw that water was everywhere in the world. He believed that the earth was a flat disk floating on a huge sea and that the motions of the sea caused earthquakes. So he suggested that the amount of water in a substance gave it its unique characteristics. Thales was the first person in recorded history to suggest a material explanation for events that others thought were the works of the gods. Another Greek philosopher,

Anaximenes, took exception to Thales' argument, insisting that air was the fundamental material. Another philosopher suggested fire as a primary material.

A century later, Empedocles of Sicily asked, "Why must there be just one basic substance?" That was too simple. He believed that water, air, fire, and earth were the four "elements," or building blocks, of matter.

The Greek philosopher Democritus believed that matter is made up of solid atoms and empty space.

THE FIRST ATOMISTS

Democritus (ca. 460–370 BC) was a Greek philosopher intensely involved in the debate on the nature of matter. Developing the ideas of his teacher, Leucippus, Democritus said that matter is made up of solid "atoms" and empty space. Materials could be broken apart because of the spaces between atoms, but if you continued to divide matter into smaller pieces, you would eventually reach a point where the remaining material could no longer be divided. At this point, the remaining solid particle would be "indivisible," or *atomos* in Greek. "Atoms" were indivisible, Democritus said, because they contained no empty space. They were the smallest pieces of substances that existed. They were not only solid, they were invisibly small. And they were all alike; that is, they were all made from the same fundamental substance. What made various materials different from each other was the shape of their atoms, or the way those atoms were packed together. Democritus's theory was beyond human experience. No one could see an atom or prove its existence. His theory was based on the assumption that a person cannot continue dividing a material object into smaller pieces forever. All the same, Democritus's

The Greek philosopher Aristotle didn't believe that atoms existed, and for centuries his views dominated European thought.

atomic theory was remarkably insightful—and 2,000 years ahead of its time.

DISSENT AND NEW ORTHODOXY

The great Greek philosopher Aristotle (384–322 BC) did not accept Democritus's theory. Aristotle preferred Empedocles' theory of the four elements, and he suggested that these elements could be transformed into one another by the degree of opposite qualities they possessed, whether they were hot or cold or wet or dry.

Aristotle is generally regarded as the greatest thinker of his age. Yet many historians believe that Aristotle's dismissal of "atomism" is the main reason that scientific inquiry into the nature of matter was blunted for nearly twenty centuries after his death. During the Middle Ages, the Catholic Church adopted Aristotle's ideas and said that it was heresy to disagree with them. The atomic theory was associated with atheism. As late as 1624, the French government issued a decree prescribing

IOÁNES
STRATENSÍS
FLANDRVS

the death penalty for anyone who disagreed with Aristotle. In the areas of politics, art, and logic, Aristotle was one of the greatest influences on Western thought, but the weight of his authority would hold back the sciences for a thousand years. During the Middle Ages, the study of matter was taken over by the alchemists.

ALCHEMY

Alchemy was the attempt to transmute, or change, one substance into another, and was therefore a direct outgrowth of Aristotle's concept of the transmutability of elements. Alchemists called the agent of transformation the "philosopher's stone," a mysterious chemical preparation that could alter the nature of substances.

Alchemists were the early pioneers of chemistry, despite the fact that today their experiments seem odd.

They toiled mightily to find this miraculous material, which would permit them to transform lead into gold. They also believed that it would help them to create an elixir of eternal youth and a medicine that would cure all human diseases.

We may smile at what today seem to us silly pursuits, yet most alchemists were learned men and women. They made some important contributions to the development of chemistry. They tested the properties of nearly every substance they could get their hands on. They did not use today's accepted scientific methods, but they did set up laboratories in which they studied some simple chemical processes, such as crystallization and distillation. Experiments of a sort were planned and carried out, and detailed records of the results were often kept. Many of the experimental techniques were first developed by Arab scientists.

The alchemists also discovered at least five true elements: arsenic, bismuth, zinc, phosphorus, and antimony. They are also credited with the discovery and description of many chemicals used in their experiments, including nitric acid, acetic acid, and ammonium chloride.

THE DAWN OF SCIENTIFIC CHEMISTRY

For centuries, alchemy dominated studies of matter. Even the great scientific figure Isaac Newton (1642–1727) is said to have spent a lot of time studying alchemy books. But other scientists, contemporaries of Newton, were engrossed in experiments that would undermine the foundations of alchemy and provide the first real evidence to support the atomic theory of matter. The first breakthroughs would come with the study of gases.

The brilliant Irish chemist Robert Boyle (1627–1691) was one such experimenter. The son of the wealthiest man in Great Britain at the time, he was well educated and was able to choose a leisurely life of scientific study, when he wasn't occupied protecting

BOYLE'S
WORKS

his estates from the depredations of the English civil war between Charles I and Oliver Cromwell. In one experiment, using a vacuum pump designed by his assistant Robert Hooke, Boyle evacuated the air in a closed chamber and proved that air was necessary for the transmission of sound and for candle flames to burn. More important, with Hooke's apparatus he had created a vacuum, something that Aristotle and others had claimed was impossible.

In his most famous experiment, Boyle sealed the opening of the shorter leg of a J-shaped glass tube and then poured mercury into the longer leg. This trapped the air in the short leg of the tube between the seal and the mercury. He continued to pour until the mercury in both the long and the short legs of the tube was at the same level. Then Boyle poured thirty more

Irish chemist Robert Boyle proved that gas pressure and volume are related, and in so doing supported the existence of atoms.

inches of mercury into the long end of the tube, in effect doubling the atmospheric pressure. The additional mercury compressed the air in the short leg of the tube to half its original volume. Thus Boyle had doubled the pressure on the trapped air and halved its volume. Boyle's law, as it is now known, proved that gas pressure and volume are inversely related. Decrease the volume of a gas and you increase its pressure, and vice versa. What was significant about this discovery was that this is the way a gas would behave if it were composed of a mass of freely moving particles with empty spaces between those particles that allowed the gas to compress or expand. This clearly supported the atomic theory of matter.

Boyle's break with the tradition of alchemy was complete. In 1661, he published *The Sceptical Chemist,* in which he directly attacked Aristotle's notion that the world was composed of the four elements of fire, earth, air, and water. He believed that substances were composed of "corpuscles" that were themselves

Clergyman Joseph Priestley produced the first carbonated beverage and is considered the inventor of soda pop.

combinations of more fundamental particles—not a bad guess at the true relationship between molecules and atoms. Boyle also discarded the alchemists' habit of secrecy and promoted the publication of all experimental data, including the results of unsuccessful experiments. In 1654, he helped to found Britain's most famous scientific body, the Royal Society.

Following Boyle, many experimenters began to study the secrets of gases, and in the process they began to map out a scientific foundation for chemistry. Joseph Priestley (1733–1804) began his career as

a clergyman in Leeds, England, but he became interested in scientific matters through his friendship with the American inventor and revolutionary Benjamin Franklin. Priestley lived not far from a brewery, and he became fascinated with the gas given off by the brewing vats; a gas later identified as carbon dioxide. He collected and experimented with quantities of this gas and discovered a way to infuse it into water, producing the first carbonated beverage and earning several awards as the inventor of soda pop.

Priestley's most important work, however, concerned his discovery of oxygen, but he misinterpreted the results of his experiment because of his belief in the "phlogiston" theory. At the time, scientists were intensely curious about the process of combustion, which they believed would reveal much about the chemical nature of substances. Scientists observed that when something burned, most of its material disappeared, leaving only fragments or ashes. They concluded that during combustion, materials gave off into the air a substance called phlogiston. A substance disintegrates into ashes when it burns because this mysterious phlogiston has been “freed” from that substance. A material like charcoal, which burned almost completely, was thought to be composed almost entirely of phlogiston.

Priestley heated mercury in air, using sunlight concentrated by a magnifying glass. The surface of the heated mercury gleamed and became coated with a red powder, which we now know to be mercury oxide. When Priestley removed this powder and heated it, it evaporated into two different gases. One was mercury vapor, which condensed quickly into droplets of pure liquid mercury. The other was a colorless gas, which Priestley found could rekindle a smoldering candle wick. Priestley even inhaled quantities of this gas and reported feeling “light and easy.” He had discovered oxygen, but at the time Priestley called it “dephlogisticated air.” He believed that in heating the mercury, it had given off its phlogiston, and the gas given off by the red powder residue was a purer kind of air without phlogiston.

Priestley was a man of unconventional religious and political views. He had supported both the American and French revolutions, and in 1791, an English mob burned down his house as a result. He eventually moved with his family to Philadelphia, where he founded the first Unitarian church in the United States.

THE FALL OF PHLOGISTON

The phlogiston theory was eventually debunked by French chemist Antoine Laurent Lavoisier (1743–1794),

French chemist Antoine Laurent Lavoisier published the first modern chemistry textbook. He described the properties of the known elements and devised a set of symbols for them.

a man who knew and admired Priestley and only with great reluctance challenged him. In 1775, Lavoisier was appointed to the National Gunpowder Commission to study the poor quality of French explosives. He moved into the Arsenal of Paris, where he was able to create a splendid laboratory. Lavoisier had a hunch that something was wrong with the phlogiston idea, and he believed that only precise and accurate measurement would reveal the truth. Other chemists had heated materials in open air. Lavoisier used a complicated, closed system of flasks and tubes that enabled him to collect and weigh all the gases and end products of his combustion experiments.

Lavoisier repeated Priestley's experiments, heating metals until they formed a "calx," a crust of what today we know is a metal oxide. For some reason, the calx always weighed more than the original unheated metal. How could the metal be giving off phlogiston if it was gaining weight? Lavoisier weighed everything he used, both before and after his experiments. He weighed the metal, the flask, and the entire closed apparatus. Lavoisier found that the entire apparatus had exactly the same weight before and after heating, lending support to what is now known as the law of the conservation of mass. In chemical reactions, no matter is destroyed, even though it may change form.

When Lavoisier weighed the calx alone, however, it was indeed heavier than the original, unheated metal. Where did the extra weight come from? Lavoisier theorized that the metal's weight gain must have come from combining with the air in the flask. He also realized that if some of the air in the flask combined with the metal, there should be a partial vacuum in the flask. When he opened the flask, Lavoisier was rewarded by a rush of air filling the vacuum that had been created. Lavoisier had demonstrated not that phlogiston flowed out of materials when they burned, but that a portion of the air was drawn into materials during combustion.

One thing still bothered Lavoisier. After measuring the vacuum that was created, he realized that only about one-fifth of the air in the flask was involved in the reaction. Why not all of it? After a discussion with Priestley, Lavoisier realized that air must consist of a combination of two gases, each with different properties, and that only one of them was involved in the reaction. Lavoisier went on to isolate this gas and

Eccentric English scientist Henry Cavendish discovered that water is made up of oxygen and hydrogen.

named it oxygen, rejecting Priestley's "dephlogisticated air." The other four-fifths Lavoisier called azote, from the Greek for "no life." Today we call this gas nitrogen.

Lavoisier's work was revolutionary. In 1787, he published the first modern chemistry textbook. In his book, Lavoisier used Boyle's definition of an element as a substance that could not be broken down into simpler substances. He carefully described the qualities of the known elements and devised a set of symbols for them and a method for describing chemical reactions. Often regarded as the father of modern chemistry, Lavoisier came to an unpleasant end. During the French Revolution, he fell afoul of the revolutionaries and died on the guillotine in 1794.

Phlogiston died a long, hard death, helped by another contemporary of Lavoisier, eccentric English scientist Henry Cavendish (1731–1810). A Cambridge University dropout, Cavendish inherited an enormous fortune and settled in London, where he lived a reclusive life as an amateur experimenter. He was terrified of women and would communicate with his female servants only through written notes. Cavendish conducted an experiment in which he added acid to metal, expecting that phlogiston would be produced by the reaction and that it would accumulate in a glass tube already filled with "dephlogisticated air," that is, oxygen. If a flame inserted into the tube continued to burn, it would prove the presence of phlogiston.

The gas produced by the reaction was not phlogiston but hydrogen, and in the presence of the flame, Cavendish noticed that droplets of liquid formed on the inner walls of the vessel. It looked like water. It tasted like water. It was water. Lavoisier was ecstatic when he heard about Cavendish's results. He repeated the experiment and announced to the world that water was a compound—a material made up of two gases chemically combined. One gas, he knew, was oxygen. He named the other gas "hydrogen," from the Greek word for "water producer." Lavoisier announced that water was an oxide of hydrogen, not one of Aristotle's elements. Thus began the era of modern chemistry.

Elements and Compounds

The process of combustion, then, did not release phlogiston, but involved the combination of oxygen with another element to form a compound, a new chemical with properties different from either of the elements that formed

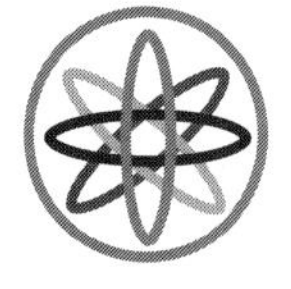

it. The discovery of oxygen and the true nature of the combustion process signaled the true beginning of the science of chemistry. Lavoisier's textbook had named the known elements and compounds, but chemistry was still in its infancy. Lavoisier had incorrectly listed heat and light among the elements, and of the other elements he listed, a number were later discovered to be compounds.

This problem plagued early chemists, who did not have the techniques to decompose all the existing compounds or to isolate pure elements. How could chemists identify and determine the nature of compounds, and upon what scientific principles were the particles that made up these compounds combining with one another? Several basic ideas were known and generally accepted by the scientific community. One such idea was the law of conservation of mass, demonstrated by Lavoisier with his experiments with oxygen in 1798. The total mass of reactants in any chemical process was always equal to the total mass of the products. That is, mass, or matter, cannot be created or destroyed.

Another general principle was first demonstrated by Joseph Louis Proust (1754–1826), a Frenchman who set up his laboratory in Spain after fleeing France to escape the guillotine. In 1799, Proust showed that no matter what processes were used to produce copper carbonate, the compound always contained the exact same proportion by weight of the elements copper, carbon, and oxygen. From this and other experiments, Proust proved the law of definite proportions. This law states that the proportions of elements in a compound do not vary. As both Priestley and Lavoisier had discovered, one unit volume of oxygen always combines with two units volume of hydrogen to form water. That was true of all chemical reactions, but it had become apparent through the study of gases. These laws seemed to suggest an underlying reality of discrete particles—atoms combining in definite and fixed ways. Could scientists describe that underlying reality?

JOHN DALTON

Dalton theorized that chemical compounds are made up of atomic particles of different weights.

John Dalton (1766–1844) was an English Quaker schoolteacher and chemist who took on this problem. Dalton had experimented with gases and the way they easily mixed with one another. He believed that this could happen only if the gases were made up of tiny individual particles with lots of empty space between them, and so he revived Democritus's name for these particles—atoms. And Dalton came to believe that all matter, not just gases, was composed of atoms. He thought of atoms as solid, indestructible spheres that had no internal structure. He believed that the atoms of different substances could only be distinguished by their different masses or weights.

In his experiments, Dalton found that different compounds could be made of different proportions, by weight, of the same elements. For example, he noticed that when he mixed three parts of carbon with four parts of oxygen, carbon monoxide always formed. But if he mixed three parts carbon with eight parts oxygen, the result was carbon dioxide. He reasoned, correctly, that a molecule of carbon monoxide is composed of one atom each of carbon and oxygen, and a molecule of carbon dioxide contains one atom of carbon and two atoms of oxygen. These results confirmed the law of definite proportions, and in 1803, Dalton came up with his own law, the law of multiple proportions, which stated that the same elements can combine in different ways to form different compounds. His experiments also demonstrated that there can be more than one atom of an element in a compound.

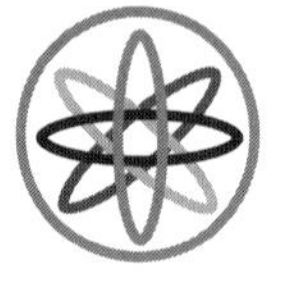

Furthermore, the proportional weights in which elements combined were always ratios of small whole numbers, like 1 to 2, or 2 to 3, or 2 to 4, as both Priestley and Lavoisier had noted. The weights of chemicals that combined were never measured as 1.3 to 2, or 3 to 4.6. The notion of whole units supported the idea that individual atoms were combining in fixed numbers and that atoms were indivisible and fundamental units.

Dalton tried to determine the relative weight of each element in compounds. He found, for example, that one gram of hydrogen always combines with exactly eight grams of oxygen to form water. From this, Dalton determined each element's "equivalent weight," that is, its weight in relation to other elements. Dalton believed that water was made up of one atom of hydrogen and one atom of oxygen. Therefore, if hydrogen, the lightest element, is given the arbitrary designation 1, the equivalent weight of oxygen is 8. Dalton continued to calculate the relative weights of other elements, and he was the first to work out a

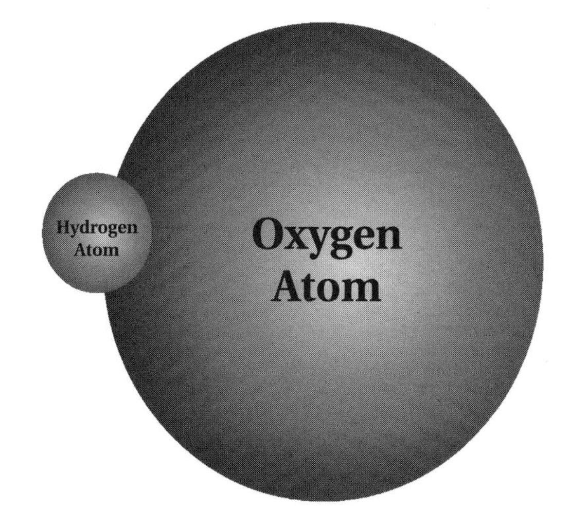

Dalton believed that water was made up of one atom of hydrogen and one atom of oxygen.

system of "atomic weights" for the known elements.

Dalton published his findings in favor of the atomic theory in a book entitled *A New System of Chemical Philosophy* in 1808. The book revolutionized the science of chemistry. With one or two exceptions, the notable scientists of the day recognized the genius of his work, but being a Quaker, he refused all recognition and ceremonial tribute. He had to be elected to the Royal Society without his own knowledge. When finally presented to the king in the scarlet robes of an Oxford scholar, a color of dress forbidden to Quakers, he tolerated the heresy only

ELEMENTS

Element	Wt.
Hydrogen.	1
Azote	5
Carbon	54
Oxygen	7
Phosphorus	9
Sulphur	13
Magnesia	20
Lime	24
Soda	28
Potash	42
Strontian	46
Barytes	68
Iron	50
Zinc	56
Copper	56
Lead	90
Silver	190
Gold	190
Platina	190
Mercury	167

because he was colorblind and saw himself as dressed in gray instead.

Dalton's views put chemistry on a sound scientific footing, but he was not right about everything. As we shall see, we have learned that atoms are not solid, indestructible spheres without internal structure. Dalton also stubbornly believed that in a free or uncombined state, elements exist only as single atoms. In Dalton's view, only compounds could contain more than one atom of an element. Because he believed that elemental oxygen is made up of single atoms of oxygen, Dalton incorrectly stated the atomic weight of oxygen to be 8 in relation to hydrogen. Many of his atomic weights were wrong because he believed that like atoms would repel each other and could not combine. But other scientists were not so sure. New

Early tables of the elements used symbols instead of letters.

experimental techniques using electricity would seriously challenge Dalton's idea.

ELECTROCHEMISTRY

In 1800, Italian scientist Alessandro Volta invented the battery and produced the first electric current. Within months of this discovery, English chemist William Nicholson used an electric current to separate the elements comprising a compound, a process called electrolysis by English physicist Michael Faraday. It was now possible to break apart many compounds and precisely measure the relative weights of the elements that composed them.

The Italian scientist Alessandro Volta invented the battery and created the first electric current.

In 1808, French chemist Joseph Gay-Lussac (1778–1850) used electrolysis to determine the exact volume of hydrogen and the exact volume of oxygen that combine to form water, rather than the weights of the elements. He found that oxygen was able to combine with precisely twice its own volume of hydrogen. Gay-Lussac hypothesized that water consists of one atom of oxygen and two atoms of hydrogen. If water contained two atoms of hydrogen rather than one, its weight relative to oxygen was only one half of Dalton's value, or to put it another way, if the atomic weight of hydrogen were set at 1, the atomic weight of oxygen had to be 16, not 8. Dalton insisted that Gay-Lussac must have made a mistake in his experiment, but Gay-Lussac's work straightened out a lot of inconsistencies in the atomic weights of the elements.

AMEDEO AVOGADRO

Studying the experiments of Gay-Lussac, Italian chemist Amedeo Avogadro (1776–1856) in 1811 put forth the idea that under identical conditions of temperature and pressure, all gases of equal volume contain an equal number of "particles." This is now known as Avogadro's law. If the electrolysis of water yields twice as much volume of hydrogen as oxygen, then water must contain twice as many hydrogen particles as oxygen particles. The formula for water could not be HO, as Dalton believed, but had to be H_2O (two atoms of hydrogen to one of oxygen).

Avogadro also experimented with hydrogen chloride (HCl). He combined one liter each of hydrogen gas and chlorine gas. He expected to get one-half liter of hydrogen chloride. He reasoned that if both hydrogen and chlorine exist as single atoms, once the atoms "pair off" only half as many "compound" atoms are left. Avogadro was amazed when his experiment yielded a full liter of hydrogen chloride. How was this possible? Avogadro suggested that if each gas particle contained two atoms of each gas, the problem was solved. One particle of hydrogen with two atoms plus one particle of chlorine with two atoms equal two particles of hydrogen chloride, each containing two

Swedish chemist Jöns Jakob Berzelius prepared a table of elements with accurate atomic weights.

atoms. Avogadro concluded that gaseous elements exist in diatomic forms—particles containing two atoms each. This refuted Dalton's "one-atom element" theory. Avogadro was the first to distinguish between atoms and what he named "molecules," and to clear up the confusion between atomic and molecular weights.

ORDERING THE ELEMENTS

By the 1820s, experiments in chemistry were revealing new elements at an amazing rate. Swedish chemist

Jöns Jakob Berzelius (1779–1848) conducted thousands of experiments on compounds to determine the exact weight ratio between their constituent elements. He is credited with the discovery of the elements cerium, selenium, silicon, and thorium. Berzelius prepared an extensive table of elements with accurate atomic weights, using a system of abbreviations based on the Latin names for the elements. This is why potassium is represented by the letter K, from the Latin *kalium*, and why gold is represented by Au, from the Latin *aurum*. Dalton had used circles with various cryptic symbols to represent elements and declared Berzelius's symbols "horrifying," but Berzelius's system proved superior and is still used today.

THE PERIODIC TABLE

By 1860, just over sixty elements had been identified, and their relative atomic weights were known. A great deal was also known about their chemical properties.

Dmitri Mendeleev created the modern periodic table of the elements in 1869.

In 1869, Dmitri Ivanovich Mendeleev (1837–1907), a chemistry professor at the University of St. Petersburg in Russia, was preparing a chemistry textbook for his students. He felt the need to put all the known elements into some kind of order. He began this monumental task by gathering every bit of information he could find about each element. He then studied and analyzed these properties looking for similarities.

Some relationships were already known and were quite suggestive. The halogens, or salt-forming elements, (fluorine, chlorine, iodine, and bromine) had to be grouped together. The alkali metals (potassium, sodium, and lithium) also had common properties. So did the metals gold, silver, and copper. Could our fundamental understanding of chemistry be advanced if we knew the reason for these relationships?

Mendeleev wrote each element on a card. He tacked each card up on the wall. For the better part of a year he studied the cards, arranging them in various rows and columns, based on what he knew about the elements' properties, searching for a logical order. At some point, Mendeleev arranged the cards in an order that struck him as meaningful. Hydrogen, as the lightest element, had been given first place in the upper left of his arrangement. Beneath hydrogen and running across the second row, he had arrayed the cards for seven elements, from lithium to fluorine, in order of increasing atomic weight. In row three, he had placed the next seven elements, from sodium to chlorine, again in order of increasing atomic weight. A pattern began to make itself evident.

Horizontally, the elements were in order of increasing atomic weight. But the elements also began to align themselves vertically in columns, and each column of elements had similar chemical properties, notably a similar valence. Valence was the term applied by chemists to describe an element's ability to combine with other elements. An element like lithium with a valence of 1 could combine with one other atom. An element with a valence of 2 could combine with two other atoms, as the oxygen atom combines with two hydrogen atoms to form water. Carbon, with a valence of 4, combines with four hydrogen atoms to form methane (CH_4). Moving across a row horizontally, the pattern of valences was 1, 2, 3, 4, 3, 2, 1. But vertically, elements with different atomic weights had the same valence, or chemical reactivity. As the mass or atomic weight of the elements increased, at regular intervals or periods elements would exhibit similar chemical properties. For this reason, Mendeleev called his chart the periodic table of the elements, and he published his first version of the table in 1869.

Mendeleev had arranged all the known elements in order of increasing atomic weight and discovered families of elements with similar chemical properties.

Other patterns revealed themselves as well, such as groupings of metals and nonmetals at opposite sides of the periodic table. In order to fit elements with similar properties into their proper columns, however, Mendeleev had to leave gaps in his table. He confidently predicted that these gaps would be filled by elements yet to be discovered, and he even predicted what chemical properties those elements would have. By 1875, chemists had discovered three new elements predicted by Mendeleev, and his reputation as the greatest chemist of his time was assured.

But what did all this mean? The elemental substances that made up the world were themselves composed of infinitesimally small atoms that differed from one another by their mass or weight, increasing in multiples of the atomic weight of hydrogen, the lightest element. But what determined their different chemical properties? Why did some atoms combine more readily than others? Why did elements combine in definite proportions? And what was the meaning of

atomic weight? Why were some atoms "bigger" than others, and bigger always in mutiples of the weight of the hydrogen atom? To learn why the elements had different chemical properties, scientists were going to have to look inside the atom. Scientists were about to embark on a journey of discovery that would overthrow some of the ideas of the early atomists and even return to the ideas of the alchemists. The atom, it turns out, is not indivisible, and in fact has a complicated internal structure. And if that internal structure is changed, one element can be transmuted into another element. As the nineteenth century drew to a close, the question of the nature of matter was taken up by physicists as well as chemists, and a new and deeper understanding of the nature of the atom was about to emerge.

The Atom Revealed

In the 1870s, English chemist William Crookes (1832–1919), investigating the effects of gas pressure on the flow of electric current, decided that he needed a sealed glass tube with a better vacuum in it than had ever been achieved before. The Crookes tube, as it came to be called, was an

To better investigate the effects of gas pressure on the flow of electric current, English chemist William Crookes invented an early version of the cathode ray tube in the 1870s.

early precursor of the cathode ray tubes found in our television sets. Crookes's device consisted of a sealed glass tube with two metal wires inside. When the wires were connected to a battery and an electric current was passed through the tube, the negatively charged wire, the cathode, gave off some kind of radiation that made the tube fluoresce, or glow. When Crookes placed a small piece of metal in front of the cathode, a shadow was cast on the glass. This indicated that the radiation was flowing in a straight line. These mysterious emissions were dubbed "cathode rays."

Scientists debated the nature of these cathode rays. The rays could cast a shadow, so some scientists asked if they were visible light, that is, were they waves? Other scientists were sure that the rays were charged particles, as Crookes himself believed, because they emerged from the negatively charged cathode. Heinrich Hertz, the discoverer of radio waves, believed that cathode rays were waves. When he passed cathode rays between two electrically charged metal plates, the cathode rays were not seen to deviate from their straight path. Since particles would have been deflected by the electromagnetic field surrounding the charged plates, Hertz believed that this experiment proved that the rays were waves. But Hertz had not counted on the speed of the rays, which were so fast that his apparatus would not have been able to detect any deflection.

In 1897, English physicist Joseph John Thomson (1856–1940) repeated the experiment that Hertz had tried earlier. Thomson, a Cambridge graduate and student of James Clerk Maxwell, the discoverer of the laws of electromagnetism, was director of the Cavendish Laboratory at Cambridge. He had earlier calculated the speed of cathode rays, and armed with this crucial knowledge and a Crookes tube with a more perfect vacuum, Thomson shot an electric current past two sheets of charged metal as Hertz had done. There was a distinct curve in the rays' path as they sped through the electromagnetic field created by the charged metal strips. This was convincing evidence that cathode rays were particles. At the suggestion of the physicist G.J. Stoney, Thomson called these particles "electrons," each carrying the smallest, indivisible quantity of electric charge.

Thomson's most revolutionary finding, however, came when he analyzed the degree of curvature in the electron's path as it passed through the electromagnetic field. Knowing the strength of the field, the speed of the particle, and the amount of its deflection, he was able to calculate the mass of a single electron. According to Thomson, that mass was only 1/1,837 that of the hydrogen atom! Thomson had discovered a particle that was smaller than the atom, which was supposedly the smallest indivisible particle that existed!

The new discovery required a revised theory of what the atom was. It was certainly not the solid, featureless ball imagined by Dalton. Thomson conceived of the atom as a sphere of some permeable or spongy material with a positive charge, and embedded in this sphere, like blueberries in a muffin or raisins in a pudding, were the negatively charged electrons. Thomson believed that electrons were always arranged in a definite way unique to each element.

RADIOACTIVITY

In 1895, German physicist Wilhelm Roentgen discovered that the electric current running through a Crookes tube produced powerful rays that could expose photographic film even if it was wrapped in a material that prevented the penetration of light. He called these rays X rays. French physicist Antoine Henri Becquerel (1852–1908) became interested in reproducing these X rays. He observed that a compound of the heavy element uranium also emitted some kind of radiation that could expose photographic film. He had not found X rays, but a completely new kind of energy that seemed to be coming from the heart of a solid substance. Marie Curie had named these emissions "radioactivity."

Back at the Cavendish Laboratory in Cambridge, the discovery of X rays and radioactivity fascinated J. J. Thomson and his bright young assistant, New Zealander Ernest Rutherford (1871–1937). Thomson gave

Rutherford was the first to suggest that alpha, beta, and gamma rays resulted from decaying atoms.

Rutherford the task of investigating the nature of this new radiation. Rutherford began by subjecting uranium emissions to a powerful magnetic field, and discovered that there were really three different kinds of rays. The path of one ray was bent by the magnet and was attracted to a negatively charged electric terminal, suggesting that this ray was positively charged and was actually a particle. Rutherford called these positively charged rays "alpha rays." He eventually realized that alpha rays were particles, and today we know them to be the nuclei of helium atoms, containing two protons and two neutrons.

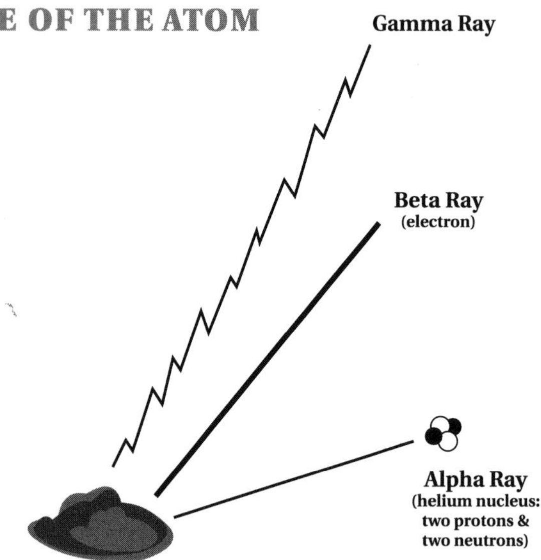

Atoms turn into other elements as they decay and release alpha and beta particles and gamma rays.

Another ray was also bent by the magnet and was attracted to a positively charged electrode. Rutherford called these rays "beta rays," and they had properties remarkably like electrons, which is what they were. The third type of ray could not be bent by the magnet. It was the strongest and most penetrating ray, very like an X ray, and was named the "gamma ray." This proved to be a true wave of electromagnetic energy. Rutherford was the first to suggest that these rays and particles were the result of the decay of atoms, which transformed themselves into other elements as they released these particles.

When Rutherford realized that the alpha ray was really some kind of massive particle, he decided to use this particle as a kind of bullet to explore the inner structure of the atom. From 1906 to 1909, Rutherford, working with his assistants Ernest Marsden and Hans Geiger, used the alpha particles from a sample of radioactive material to bombard very thin sheets of metal foil. On the other side of the metal foil was a photographic plate to record the passage of any alpha particles that passed through the metal. If J. J. Thomson's model of the atom was correct, the expectation was that almost all of the heavy alpha particles would travel straight through the metal foil, pushing aside the tiny electrons or being only slightly deflected by their negative charge.

When his assistants performed the experiment, they thought that they had made a mistake. Most of the alpha particles went straight through the metal foil, but a very few were deflected and scattered in odd directions, and a few particles were even deflected at angles greater than ninety degrees, back toward the source of the radiation. Rutherford was astounded and stated, "It was almost as incredible as if you had fired a fifteen-inch shell at a piece of tissue paper and it came back and hit you."

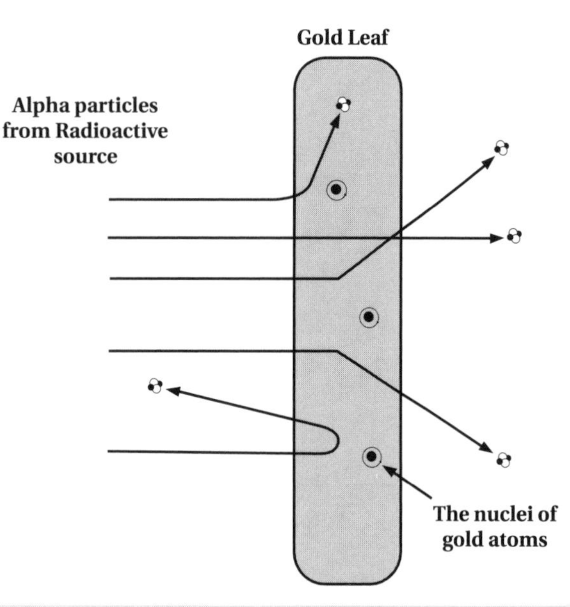

Rutherford's scattering experiment showed that the atom is mostly empty space surrounding a nucleus.

Rutherford suggested that some electrical force powerful enough to deflect the huge particles had to be involved, and that this force was concentrated within an extremely small space because only those few alpha particles that came near this force were diverted by it. In Rutherford's words, "This scattering backwards must be the result of a single collision and, to have a deflection of this magnitude, the greater part of the mass of the atom must be concentrated in a minute nucleus carrying a charge." And because most of the alpha particles had passed right through the metal foil, Rutherford proposed that an atom is, in fact, mostly empty space, with an extremely dense core—or nucleus—surrounded by empty space and electrons. Rutherford had introduced the modern idea of the nuclear atom. The notion that electrons surround, or "orbit," the nucleus, like a miniature solar system, was a powerful analogy, and Rutherford's model came to be known as the planetary model of the atom. The electrons were bound to the nucleus by the attraction of their opposite electric charges, the way that gravity binds the planets to the sun.

In later experiments, by bombarding the atoms of gases with alpha particles, Rutherford was able to demonstrate the presence of distinct positively charged particles within the nucleus. Rutherford called these positively charged nuclear particles "protons."

Rutherford also determined that the electrons were the only particles involved in chemical reactions and that the composition of the nucleus was not altered in chemical reactions. When he bombarded nitrogen atoms with alpha particles, however, he was able to transmute nitrogen into oxygen by changing the number of protons in the nucleus, achieving the dream of the alchemists. Rutherford was awarded the Nobel Prize for his work in 1908.

ATOMIC NUMBER

The nuclear atom was a neat little trick, but there were still a number of unresolved questions. One such problem was the discrepancy between atomic weight and the atom's electric charge. In 1913, British physicist Henry Mosely (1887–1915) decided to bombard a number of different metals with electrons and measure the frequency of the X rays the metals produced. He first thought that the frequency of the X rays increased when they were emitted by elements with greater atomic weight. But several discrepancies in his data made him realize that the X-ray frequency was increasing as the electric charge of the nucleus increased. Mosely called the units of positive electric charge in the nucleus the "atomic number," to distinguish them from atomic weight. Now the workings of the electromagnetic force that held the atom together were becoming clearer. The positive electric charge in the nucleus was carried by the protons, and the more protons, the greater the electric charge. The number of positively charged protons in the nucleus was matched for each atom in a normal state by the number of orbiting negatively charged electrons.

Mosely died fighting at Gallipoli, Turkey, in World War I, but his findings helped scientists to improve the arrangement of the elements in the

periodic table according to their atomic numbers, rather than just their atomic weights. The concept of atomic number clarified the connection between an element's protons, electrons, and behavior in chemical reactions.

But there was still a problem. Why was there this discrepancy between atomic number and atomic weight? Why were atoms so much heavier than the number of protons their nucleus contained? The nucleus of the oxygen atom, for example, had an electric charge eight times the charge on the hydrogen nucleus because it contained eight protons. But its atomic weight was sixteen times the weight of the hydrogen nucleus. What made up this extra mass? And if the nucleus of the atom was made up of positively charged protons, why didn't the nucleus fly apart as a result of the repulsive force of particles with the same electric charge? What held the nucleus together? What other mysteries did the nucleus of the atom hold?

THE NEUTRON

If you look at a modern periodic table, the atomic weights of the elements are at the bottom of each element's box. The atomic weight of chlorine, for example, is given as 35.453. But if atomic weights are supposed to be whole multiples of the weight of the hydrogen atom, how can an element have a fractional atomic weight like this? It turns out that this number is an average of the atomic weights of two different types of chlorine atoms found in a sample of the substance. J. J. Thomson had discovered two forms of neon gas with different atomic weights (neon-20 and neon-22), but he could not account for the mass difference.

English chemist Frederick Soddy (1877–1956), who had worked with Rutherford, coined the term "isotope" to describe these variations of the basic elements. Soddy studied the radioactive decay of the element thorium and noted that it transmuted itself into different versions of thorium with the same chemical properties but different atomic weights. Because their atomic numbers and chemical properties were identical, Soddy believed that all of these variations should occupy the same place in the periodic table. Because each isotope had the same number of protons in its nucleus, its electrical and chemical characteristics were the same, but its mass differed from other isotopes. This discrepancy disturbed scientists and threatened to invalidate the elegant organization of Mendeleev's periodic table. How could a single element have different atomic weights?

The answer did not begin to emerge until the 1930s, with the work of English physicist James Chadwick (1891–1974), another student of Rutherford's. After the work of Thomson and Rutherford, and the discoveries of the electron and the proton, many scientists suspected that the "extra" mass in the atomic nucleus was caused by the presence of an as-yet-undiscovered particle with about the same mass as the proton. The problem was that this particle would lack either a positive or negative electric charge. It was electrically neutral and would not be influenced or deflected from its path by an electric or magnetic field. If such a particle existed, it could not be detected by the usual means. Chadwick solved this problem. In 1932, he bombarded the metal beryllium with alpha particles. He placed a piece of paraffin on the opposite side of the beryllium sample. No radiation could be detected between the beryllium and the paraffin, but Chadwick was able to detect protons being ejected from the paraffin with great force. Here was evidence that an electrically neutral particle, massive enough to push protons around, was emanating from the beryllium. Chadwick called the particle a "neutron."

Now everything began to fit into place. Chadwick's discovery resolved the discrepancy between atomic weight and atomic number. The number of protons in an element always stayed the same and determined the number of electrons bound to the atom in its normal state, and hence the atom's chemical properties. But the total mass of an atom was also determined by the number of neutrons in the nucleus. These neutrons added "weight" but no electric charge. An element might have different numbers of neutrons in its nucleus, and this determined which isotope it was. The nucleus of chlorine-35, for example, contained 17 protons and 18 neutrons, but chlorine-37 contained 17 protons and 20 neutrons.

An explanation for what held the nucleus together against the repulsive force of the tightly packed protons was not long in coming. In 1935, Japanese scientist Hideki Yukawa (1907–1981) proposed the existence of a "nuclear force" exerted by all protons and neutrons. It could exert its effects only over a very short range, the diameter of the nucleus, but it was a very powerful force that could overcome the electric repulsion of the protons and hold the nucleus together. The neutrons in the nucleus, exerting nuclear force but no electric charge, helped to keep the whole structure together. Yukawa's theory was based on the idea that the nuclear force resided in a "transfer particle," about one-tenth the mass of a proton or neutron, that rapidly shuttled back and forth between the larger nuclear particles, binding them together. In 1947, the English physicist Cecil Frank Powell (1903–1969) detected such a particle, which was named a pi-meson, or pion. Yukawa's theory was proven, and he received the Nobel Prize in 1949.

The picture of the atom now seemed to be complete. Unlike Dalton's solid little ball, the atom had turned out to have a very complex structure of subatomic particles. The new model was an elegant one in which the atom functioned like a tiny solar system. It was, unfortunately, as many scientists realized, a model that wouldn't quite work. Though this picture of the atom is useful and is still taught today, it is not a complete picture. If the electrons were orbiting the nucleus of the atom, Maxwell's theory of electromagnetism required them to give off energy. All electrically charged particles that accelerated, that changed their speed or direction, as orbiting electrons had to do, had to give off energy. But if electrons gave off energy, and all they had to begin with was their kinetic energy of motion, they would have to slow down, losing momentum and spiraling into the nucleus. The whole atomic structure would collapse. The Rutherford model of the atom was a great scientific achievement, the result of painstaking experiments on objects too small to actually observe and ingenious theorizing about the meaning of the experimental results. Now all that work was in jeopardy because the model could not work.

The World of the Quantum

The problem of the stability of electron orbits in the Rutherford model of the atom found its solution when Danish physicist Niels Böhr (1885–1962) began to think about the atom in terms of some new ideas that had been developed by Max Planck (1858–1947) and Albert

Einstein (1879–1955) at the very beginning of the twentieth century.

Planck, working at the University of Berlin, was trying to solve an odd problem in the way objects radiated energy when they were heated. The frequency distribution of the radiated energy could not be explained by the laws of classical physics that conceived of light as made up of continuous waves. Unable to find a more acceptable explanation, in 1900, in what he described as "an act of desperation," Planck announced his new "quantum theory." The theory states that objects emit energy in tiny discrete packets or bundles, rather than as continuous waves, but the "size" or amount of energy in the packet is related to the frequency detected when the wavelike qualities of energy are measured. Planck called these tiny packets of energy "quanta." A single bundle of energy is called a "quantum." Planck suggested that objects emit and absorb energy only in whole quanta, never in fractional amounts of energy.

In 1905, Albert Einstein interpreted the results of another experiment in a way that confirmed the quantum theory.

Niels Böhr had worked under J. J. Thomson at Cambridge and Ernest Rutherford in Manchester before returning to Copenhagen in 1916. Planck's theory of the quantum intrigued him, and he began to see a relationship between these little bundles of energy and the way electrons might behave within the atom. Böhr rejected Rutherford's model of the atom as a miniature solar system and said that electrons don't emit energy as they orbit the nucleus because they are not really orbiting in the sense that scientists had previously understood the term. It was more useful to think of these orbits as "energy levels." In a fixed orbit or energy level, the electron is in a stable state and emits no energy. But when an electron absorbs a quantum of energy, it jumps to a "higher" orbit or energy level, and is said to be in an excited state. When the electron jumps back to its original energy level, it

Niels Böhr discovered that when an electron absorbs a quanta of energy, it jumps to a higher orbit or energy level. When it jumps back to its original energy level, the electron emits energy in the form of a photon.

emits energy in the form of a "photon," which is a quanta, or unit of energy in the range of visible light. The energy of the quantum emitted when the electron returns to its original state is exactly the same as the amount of energy the electron absorbed to jump to a higher orbit.

In 1923, Frenchman Louis de Broglie (1892–1987), studying Einstein's proposition that matter and energy were equivalent, suggested that electrons, while having some properties of particles, were also composed of "matter waves." Following up on this idea, Austrian physicist Erwin Schrödinger (1887–1961) said that the electron did not so much orbit the nucleus as oscillate around it in the form of a "standing wave." A standing wave is a wave that waves but doesn't move through space. An example of a standing wave is the waving jump rope held by two children. This standing wave has a particular wavelength, and the electron can jump only to orbits or energy levels that are whole multiples of this wavelength, not fractions. The electron can fit into an orbit that allows it to oscillate back and forth once, or twice, or three times, but not one and a half times. That odd-sized orbit is simply not permitted by nature. Since electrons could shift only to orbits of certain sizes, Schrödinger's idea helped to explain why energy was emitted by

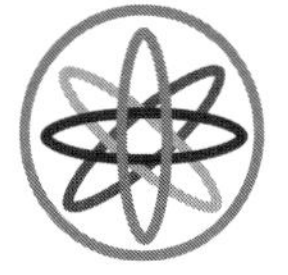

electrons only in certain size quanta. Electrons radiated photons of light only when they shifted orbits or energy levels, and then the energy of the photon depended on the type of shift experienced by the electron.

Böhr , de Broglie, and Schrödinger had created a model of the Rutherford atom that was stable. Unfortunately, the explanation forced scientists to abandon any common sense picture of what the atom might be like. The electron, rather than orbiting the nucleus like a little satellite, existed as a kind of ghostlike presence, a wave-particle, inside the atom. What it really was became increasingly difficult to describe outside the realm of mathematics. But the theory explained so much about the structure of the atom, how energy is transmitted, and how chemical reactions took place, that it was difficult to reject. The new conception of the atom was to become even stranger.

VIRTUAL PARTICLES

One of the remaining problems facing scientists had to do with the strong nuclear force and the "exchange particles" that carried this force, the pi-mesons that supposedly bound the protons and neutrons of the nucleus together. Such particles were somehow created in the nucleus and quickly shuttled back and forth between the nuclear particles they held together. Hideki Yukawa had calculated that one of these particles had to be about 270 times as massive as an electron. But how could such a massive particle just appear from nowhere? Its existence violated the principle that neither matter nor energy could be created or destroyed.

The answer came from the innovative work of German physicist Werner Heisenberg (1901–1976). Heisenberg had studied the conclusions of de Broglie and Schrödinger about the nature of the electron, and in 1927 he set forth his "uncertainty principle." The uncertainty principle states that you cannot simultaneously measure both the position and the momentum of an object with complete accuracy. With large objects, the inaccuracy is very small. But on the scale of subatomic particles, and the electron in particular, the inaccuracy is significant. This physical limit on obtaining precise information about the movement of subatomic particles indicated to Heisenberg that the world of the atom could not be explained in the same terms as the everyday world, and he abandoned any attempt to devise a visual model for what the atom was. The precise "orbits" of electrons could never be measured. This had profound implications for the way scientists viewed the world, but it also solved the problem of the exchange particles.

As a result of relationships Einstein had discovered, it was possible to substitute energy and time for position and momentum in Heisenberg's uncertainty formula. That is to say, at any exact point in time the energy (or mass, according to Einstein) of an object remains uncertain. If the time interval was small enough, so small in fact that it was impossible to measure, particles could come into being, violating the laws of conservation of mass and energy, as long as they disappeared again before they could be detected! All things were possible if they occurred between the ticks of the clock, so that their impossibilities could not be measured. Because such particles could not exist in real time in the real world, they were called "virtual particles." As bizarre as the notion of virtual particles like mesons may seem, they are real, and in fact they have been detected. Within the atom in its normal state, they are undetectable, but if enough energy is added to the atom from an outside source, mesons can be created and can exist as independent particles without violating the laws of conservation of matter and energy. Bombarding atoms with enormous amounts of energy in particle accelerators has proven the key to the discovery of many new particles that would not otherwise show themselves.

QUARKS AND GLUONS

In fact, by the mid-1960s, so many new particles had been discovered that physicists were forced to abandon the older model of the atom as composed of such ordinary subatomic particles as protons, neutrons, and electrons. It no longer seemed that these particles were truly fundamental. Physicists began to suspect that a more basic form of matter composed all these particles.

In 1963, American theoretical physicist Murray Gell-Mann (1929–) proposed that all subatomic particles were themselves composed of even smaller particles, which he called "quarks," after a phrase in James Joyce's novel *Finnegan's Wake.* Gell-Mann stated that quarks were the truly fundamental building blocks of matter. His theory required six of them, though nobody could say for sure what qualities made them different from each other. So they were given odd names: up, down, charm, strange, top, and bottom. Quarks were very odd particles indeed. They had fractional electric charges, only one-third or two-thirds of the electric charge on the proton or electron. More massive particles, like protons and neutrons, were combinations of three quarks. It was the quarks that were held together by the strong nuclear force, so much so that they could never be pulled apart in order to be observed in isolation. The exchange particles that carry the nuclear force and shuttle back and forth between bound quarks were called "gluons," and they, too, are virtual particles that cannot exist in real time.

As scientists built more powerful particle accelerators, they were able to discover evidence for the existence of quarks, confirming Gell-Mann's work. The quarks themselves can never be detected, but

Enrico Fermi (left) *and Walter Zinn posed for this photo while collaborating on developing the first controlled nuclear reaction at Argonne Laboratory in Illinois, in August 1946.*

they quickly decay into other particles that can be detected. In 1995, scientists at the Fermi National Accelerator Laboratory near Chicago finally found evidence for the existence of the top quark, the last of the six quarks to reveal itself. The discovery of these quarks was proof of what physicists now call the Standard Model of atomic structure.

In the Standard Model, all the more massive particles are made up of combinations of two or three quarks. Less massive particles, like electrons, mesons, and neutrinos, are not composed of quarks and are considered fundamental particles in their own right—a family of particles known as leptons. The various fundamental forces of nature that bind these particles together are all carried by exchange particles. The electromagnetic force is transmitted by the photon. The strong nuclear force is transmitted by the gluon, and the weak nuclear force, required to explain radioactive decay, is carried by particles known as W and Z bosons.

The discovery of new particles continues. In 1928, English physicist Paul Dirac (1902–1984) predicted the existence of a positively charged electron, known as an antielectron or "positron." Positrons were detected in 1936. Positrons are identical to electrons in every way, except for an opposite electric charge. Antiprotons were discovered in the late 1950s. Antineutrons were discovered in 1960. Today, physicists know that every particle

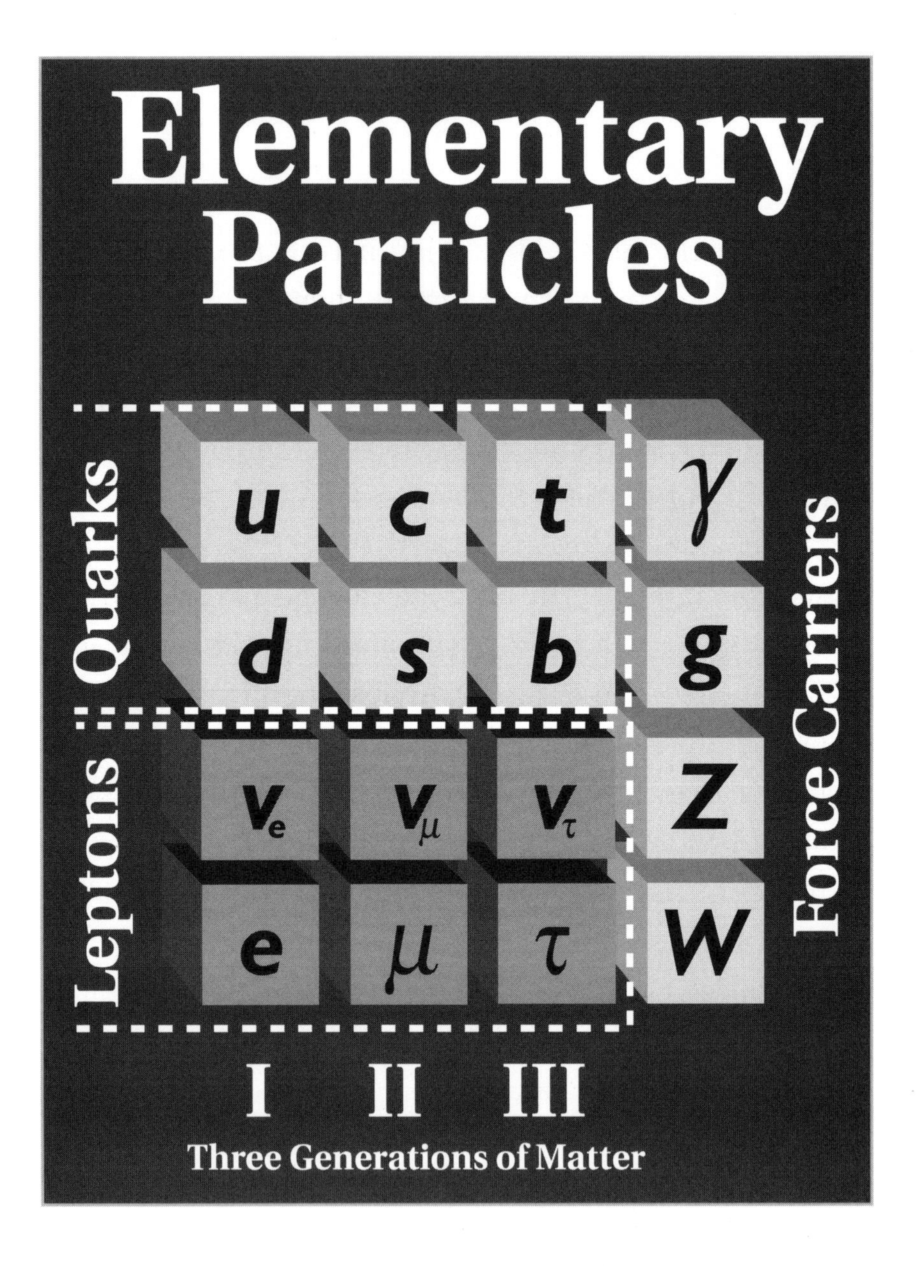
Elementary Particles
Quarks
u c t
d s b
Leptons
ν_e ν_μ ν_τ
e μ τ
γ
g
Z
W
Force Carriers
I II III
Three Generations of Matter

has an antiparticle, even the tiny neutrinos. When a particle and its antiparticle collide, they annihilate each other in a bust of energy. Theoretically, nature does not play favorites, and there should be as many antiparticles in the universe as ordinary particles, but this does not appear to be the case. Scientists wonder why.

We have come a long way from the tiny, indivisible spheres of Democritus and the early atomists. We now know that the atom has a complex internal structure of subatomic particles, and that those particles obey laws of nature very different from those we are familiar with in our everyday experience. Our current model of the atom allows us to explain many things: how the fundamental elements differ from one another, how chemical combinations occur, why radioactive elements decay, how objects radiate energy, and how to build

This is the Standard Model of subatomic particles.

everything from televisions to lasers. It has also enabled the development of the atom bomb and other weapons of mass destruction.

There is also much that we don't yet know. Can we ever describe the reality behind subatomic objects that sometimes behave like particles and sometimes like waves? Will we ever discover the relationship between the three forces that govern atomic behavior and the fourth force in the universe, gravity, thereby achieving the long sought after goal of a "theory of everything"? Will we find a deeper reality behind the quantum laws of probability that make electrons and other particles seem to live a ghostly, unreal existence? Are subatomic particles stable, or, as some scientists believe, will protons decay in billions of years, bringing an end to the material world as we know it? And the greatest cosmic question of all, how did all of these particles come into existence? As far as we have come, there is certainly a lot more to discover for the next generation of young scientists.

Glossary

alchemy A medieval practice in which people attempted to transform one element into another, especially lead into gold.

alpha particle The nucleus of a helium atom, containing two protons and two neutrons.

antiparticle Every particle has an antiparticle with the same mass but an opposite electric charge. When a particle and its antiparticle come into contact, they annihilate each other.

atom The basic unit of an element, consisting of nuclear particles and electrons (and their constituent particles, such as quarks and carrier particles). Most of the atom is empty space.

atomic number The number of protons in an atom's nucleus.

atomic weight The total number of protons and neutrons in an atom's nucleus; also called atomic mass.

beta particle An electron; beta radiation is a high-speed stream of electrons.

cathode ray The stream of electrons emitted by the negative electrode in a vacuum tube (the Crookes tube).

compound Two or more elements chemically bound together in a fixed proportion.

electrolysis A process in which an electric current is used to separate elements in a compound.

electron The negatively charged particle in an atom. Electrons are now known to have properties of both particles and waves, and "clouds" of electrons "orbit" the nucleus.

element In modern physics and chemistry, any substance that cannot be broken down into different substances by ordinary chemical means.

gamma rays A stream of high-energy photons emitted as radiation from an atom's nucleus.

gluon A particle that carries, or transmits, the strong force in the nucleus of an atom.

gravity The attractive force exerted by every object on other objects. The "graviton" is believed to be the particle that carries the force of gravity, but so far it has not been detected.

Heisenberg's uncertainty principle A principle guiding the quantum world which states that it is impossible to measure simultaneously and accurately both the energy and time (or location and momentum) of subatomic particles.

ion An atom with an electric charge, usually caused by the removal or addition of outer electrons.

isotopes Atoms of the same element that have different numbers of neutrons in the nucleus and thus have different atomic weights.

mass The amount of matter an object contains, or the degree of an object's resistance to being disturbed (its inertia). Though not precisely accurate, weight is a term commonly substituted for mass.

meson A subatomic particle that is made up of two quarks; for example, a pion.

molecule The smallest unit of any compound in which two or more atoms are bound together. Bonded atoms can be of the same element (as in a molecule of any gas) or of different elements (as in water).

neutron A subatomic particle in the nucleus that has no charge and is made up of three quarks.

nuclear decay A process in which an atomic nucleus becomes less massive by splitting apart or giving off particles as radiation.

nucleus The central part of the atom that contains most of the atom's mass; consisting of protons and neutrons.

particle decay A process in which a fundamental particle transforms itself into a completely different fundamental particle.

periodic table Table of the elements organized by atomic number and atomic weight, as well as by similar chemical properties.

phlogiston A mysterious material thought to be associated with burning. When a substance burned, alchemists believed phlogiston was freed from it. When heat caused a substance to solidify, they thought phlogiston had been added to it.

photon A particle of light; the carrier of the electromagnetic force.

positron The antiparticle of an electron. It is like an electron (with the same mass), but with a positive electric charge.

proton The positively charged particle in the atomic nucleus.

quantum In quantum theory, minute "packages" of light energy (plural, quanta).

quark The fundamental particle that makes up protons and neutrons.

radioactivity Spontaneous disintegration of an atom's nucleus to form a different nucleus, during which particles and energy are emitted.

Standard Model The current model of the atom that states that all atoms are made up of six quarks and six leptons, acted on by four forces that are carried by exchange particles.

strong force The force that holds the atomic nucleus together. Gluons transmit the strong force among the quarks that make up subatomic particles in the nucleus.

valence The ability of an element to combine with other elements through the electrons in its outer orbit. Valence determines the chemical properties of elements.

virtual particle A very short-lived particle that would violate the law of conservation of energy and mass if it could be detected.

X rays Extremely energetic electromagnetic rays that easily penetrate most materials, except platinum and lead, and that expose photographic film.

For More Information

Web Sites

Chemistry

Alchemy

http://www.levity.com/alchemy/home.html

This site offers history, pictures, resources, bibliography, and readings—just about everything you might want to know about alchemy and alchemists.

Atomic Structure and Matter: Atoms, Molecules, and Compounds

http://www.ill.fr/dif/3D-crystals/index.html

A visual guide to atoms, molecules, and compounds. For high school level and higher.

Basic Chemistry
http://www.chem4kids.com
This site provides a basic overview of chemistry.

Discovery of the Elements
http://smallfry.dmu.ac.uk
Search the site to find the history of discovery for the elements in the periodic table.

History of Chemistry
http://www.woodrow.org/teachers/ci/1992
History of chemistry by scientist or subject.

A Visual Interpretation of the Table of Elements
http://www.chemsoc.org/viselements
Click on any element in the periodic table to view its structure and to access a wealth of information about it.

PHYSICS

The Amazing World of Electrons and Photons
http://library.thinkquest.org/16468/gather/english.htm

The Big Bang
http://hepwww.rl.ac.uk/pub/bigbang/part1.html
The origins of the universe explained.

Chemical Elements
http://www.chemicalelements.com

Early Atomic Theories
http://wine1.sb.fsu.edu/chm1045/notes/Atoms/AtomStr1/Atoms02.htm
Overview of the evolution of early understanding of atomic structure (through J. J. Thomson and Ernest Rutherford).

Fermilab

http://www.fnal.gov/directorate/public_affairs

The site for Fermilab contains basic science history and information on the atom, quarks, the Standard Model, the discovery of new particles (especially the neutrino), and other interesting atomic news and information.

Lawrence Berkeley Lab

http://www.lbl.gov

The Web site for Lawrence Berkeley Labs; contains information on atom research, including the discovery of new elements. Also offers a periodic table of the elements and a periodic table of isotopes.

Laws of Physics

http://www.alcyone.com/max/physics/laws/index.html

This site defines the laws of physics.

Life, the Universe, and the Electron

http://www.iop.org/Physics/Electron/Exhibition

A fine site celebrating the 100th anniversary of the discovery of the electron.

Nuclear Fusion

http://www.pppl.gov/fusion_basics/pages/fusion_basics.html

A fine site for learning all about nuclear fusion, from the Princeton Plasma Physics Lab.

Particle Accelerators

http://www2.slac.stanford.edu/vvc/home.html

The site of the Stanford Linear Accelerator Center. Take a virtual tour of an atom-smasher.

The Particle Adventure

http://ParticleAdventure.org

Overview of the discovery of new particles and of the Standard Model of the atom.

Physics 2000

http://www.colorado.edu/physics/2000

Current issues and discoveries in physics from the University of Colorado.

WebElements Periodic Table

http://www.webelements.com

For Further Reading

Asimov, Isaac. *Asimov on Chemistry: A Journey into the Land of the Chemical Elements.* New York: Doubleday, 1974.

Asimov, Isaac. *Atom: Journey Across the Subatomic Cosmos.* New York: Dutton, 1991.

Atkins, P. W. *The Periodic Kingdom.* New York: Basic Books, 1995.

Emsley, John. *Molecules at an Exhibition: The Chemistry of Everyday Life.* Los Angeles: Getty Center for Education in the Arts, 1999.

FOR FURTHER READING

Fox, Karen. *The Chain Reaction: Pioneers of Nuclear Science.* New York: Franklin Watts, 1998.

Henderson, Harry. *Nuclear Physics.* New York: Facts on File, 1998.

Read, John. *From Alchemy to Chemistry.* Mineola, NY: Dover, 1995.

Stwertka, Albert. *A Guide to the Elements.* Rev. ed. New York: Oxford University Press, 1999.

Stwertka, Albert. *The World of Atoms and Quarks.* New York: Twenty-First Century Books, 1995.

Topp, Patricia. *This Strange Quantum World & You.* Nevada City, CA: Blue Dolphin, 1999.

Yount, Lisa. *Antoine Lavoisier: Founder of Modern Chemistry.* Springfield, NJ: Enslow, 1997.

Index

Credits

ABOUT THE AUTHOR

For more than ten years, Natalie Goldstein has been writing educational materials about the environmental and life sciences. She has worked for the Nature Conservancy, the Hudson River Foundation, the Rare Animal Relief Effort (now part of the World Wildlife Fund), and the Audubon Society. She also does volunteer work for the Rainforest Alliance and the Sierra Club. A member of the National Association of Science Writers, the Society of Environmental Journalists, and the Editorial Freelancers Association, Ms. Goldstein holds master's degrees in environmental science and education.

PHOTO CREDITS

Cover © Patrice Loiez, Cern/Science Photo Library; cover inset © Mehau Kulyk/Science Photo Library; pp. 8, 14, 23, 27, 36, 48, 54 © Northwind Picture Archives; pp. 10, 12, 20, 30, 40, 43 © Bettmann/Corbis; p. 16 © David Lees/Corbis; pp. 46, 84 © Archive Photos; p. 59 © Deutsche Presse Agentur/Archive Photos; p. 76 © AP Wide World. Diagrams on pp. 39, 60, 62, 86 by Geri Giordano.

DESIGN AND LAYOUT

Evelyn Horovicz